George Cary Comstock

Studies in Spherical and Practical Astronomy

George Cary Comstock

Studies in Spherical and Practical Astronomy

ISBN/EAN: 9783337395940

Printed in Europe, USA, Canada, Australia, Japan

Cover: Foto ©berggeist007 / pixelio.de

More available books at **www.hansebooks.com**

BULLETIN OF THE UNIVERSITY OF WISCONSIN

SCIENCE SERIES, VOL. 1, NO. 3, PP. 57–107.

STUDIES IN SPHERICAL AND PRACTICAL ASTRONOMY

BY

GEORGE C. COMSTOCK

Director of the Washburn Observatory

PUBLISHED BY AUTHORITY OF LAW AND WITH THE APPROVAL OF THE REGENTS OF THE UNIVERSITY

MADISON, WIS.
PUBLISHED BY THE UNIVERSITY
JUNE, 1895

PRICE 40 CENTS

Studies in Spherical and Practical Astronomy - Comstock

BULLETIN OF THE UNIVERSITY OF WISCONSIN
SCIENCE SERIES, VOL. 1, NO. 3, PP. 57—107, JUNE, 1895

STUDIES IN SPHERICAL AND PRACTICAL ASTRONOMY.

BY GEORGE C. COMSTOCK,
Director of the Washburn Observatory.

The following pages contain an exposition of methods for the treatment of certain problems in spherical and practical astronomy, which, from his own experience, the author has found to be advantageous in practice. For the most part these methods are original and hitherto unpublished, but in part they are due to others, whose published exposition of them is not readily accessible to American students. In cases of the latter kind due acknowledgement is made in connection with the presentation of the subject matter, but I have not scrupled to modify or to completely alter the mode of presentation of those subjects which have been treated by others, adopting in each case that method which has seemed to me simplest and most easily followed.

MINOR SUGGESTINOS.

The Reduction of Level Readings.—To determine the inclination of a nearly horizontal line or plane by use of a spirit level, Chauvenet[1] gives rules which in all cases require the same operations to be performed with the level, but in which the mode of treatment of the level readings depends upon the manner in which the scale is graduated, one method when the zero is at the end of the scale and another when it is in the middle of the scale. The modes of reduction are sufficiently illustrated in the following examples given by Chauvenet.[2]

[1] Spherical and Practical Astronomy, Vol. II, §§ 52, 55.
[2] *Loc. cit.*

Zero at end.		Zero in middle.	
W.	E.	W.	E.
29.1	31.2	+ 64.0	+ 13.5
35.4	24.9	− 10.1	− 60.7
64.5	56.1	+ 77.5	
56.1		− 70.8	
$z = 8.4 \div 4 = 2.1$ *div.*		$z = +6.7 \div 4 = +1.675$ *div.*	

A method of reduction which is the same for both types of level, and which is in most cases more convenient than the above, is as follows: In the square array of numbers which constitute the observed readings of the level, take the diagonal differences. The mean of the two diagonal differences is the inclination of the line in half divisions of the level. That end of the line is the higher which is adjacent to the greatest single reading. If the level readings have been correctly made the two diagonal differences will be the same, and the reduction thus serves as a check upon the accuracy of the record.

Thus, from the readings given above, we see at a glance that in the first case $z = 4.2$ half divisions and the readings have been correctly made. In the second case $z = 3.35$ half divisions and there is a discrepancy of 0.1 *div.* in the readings.

Although I cannot doubt that this simple mode of reducing level readings has been employed by others, I do not find it in any of the treatises upon practical astronomy to which I have access.

To Focus a Telescope.—Let the telescope be directed to a circum-polar star near culmination and introduce between the objective and the star an opaque screen pierced with a circular aperture from 10 to 20 *mm* in diameter. As the aperture is moved about in front of the objective an image of the star will be formed by different parts of the objective, and if the telescope is not perfectly focused these images will fall at slightly different parts of the field; e. g., let the aperture be held opposite the upper part of the objective and the star's image be bisected with a horizontal thread. Then shift the aperture to the lowest part of the objective and note whether the image of the star is sensibly displaced from the thread. If the image moves in the same direction with the aperture in the screen, the eye end should be drawn out; if in the opposite direction it should be pushed in until a position is found at which there is no displacement of the star image.

By this process the telescope may be so adjusted that the error of focusing shall not exceed 1:10000 part of the focal length, provided it is so firmly supported as to be free from the effect of accidental tremors and vibrations, e. g. the telescope of a transit instrument.

I.—A SIMPLE BUT ACCURATE EXPRESSION FOR THE ATMOSPHERIC REFRACTION.

Bessel's expression for the refraction[1]

$$R = \alpha \beta^{A} \gamma^{\lambda} \tan z$$

is commonly employed for all accurate computations of the refraction, and when so employed requires that the five quantities, $\alpha, \beta, \gamma, A, \lambda$, shall be interpolated from specially prepared refraction tables. It is the purpose of the present paper to so transform this expression that the refraction may be computed without recourse to these tables.

Since the refraction admits of development in terms of the odd powers of *tan z*, we may write for the mean refraction:

$$R_m = \alpha \tan z = \alpha_1 \tan z - \alpha_3 \tan^3 z \ldots \text{etc.}$$

$$= \alpha_1 \left(1 - \frac{\alpha_3}{\alpha_1} \tan^2 z\right) \tan z \quad (\text{approximately})$$

The Pulkowa Refraction Tables are presumably the most accurate ones available at the present time, and from these tables I find:

$$\alpha_1 = 57''.584 \qquad \alpha_3 = 0''.0640$$

If with these values we compute

$$\alpha = \alpha_1 \left(1 - \frac{\alpha_3}{\alpha_1} \tan^2 z\right)$$

and compare it with the tabular values of α we shall find the following satisfactory agreement:

z	0°	20°	40°	60°	75°
	"	"	"	"	"
Tabular α	57.586	57.577	57.538	57.386	56.694
Formula	57.584	57.576	57.537	57.391	56.693

The quantity λ is a complicated function of the zenith

[1] *Tab. Reg.*, LXII.

distance, z, but for values of z less than 75° it may be represented by the empirical formula:

$$\lambda = 1 + h \tan^2 z \qquad h = 0.001362$$

The following comparison shows the degree of accuracy with which this formula represents the tabular numbers:

z	50°	60°	70°	75°
Tabular λ	1.0022	1.0044	1.0103	1.0188
Formula	1.0019	1.0040	1.0103	1.0190

If we represent by ε the adopted coefficient of expansion of air per degree C., by τ_0 the normal temperature of the refraction tables, and by τ any other temperature, we shall have:[1]

$$\gamma^{\lambda} = \left[1 + \varepsilon(\tau - \tau_0)\right]^{-\lambda}$$

Developing this expression by means of the exponential series it becomes, when the terms of the order ε^2 are neglected,

$$\gamma^{\lambda} = \frac{\varepsilon^{-1} + \tau_0}{\varepsilon^{-1} + \tau}\left\{1 - \log_e\left[1 + \varepsilon(\tau - \tau_0)\right] h \tan^2 z\right\}$$

$$= \frac{\varepsilon^{-1} + \tau_0}{\varepsilon^{-1} + \tau}\left\{1 - \varepsilon h \tan^2 z (\tau - \tau_0)\right\}$$

For zenith distances less than 75° the exponent A does not sensibly differ from unity, and we have

$$\beta^{A} = \frac{B}{B_0}$$

where B_0 is the normal barometric pressure of the tables and B is the actual pressure at any time, *i. e.* the reading of the barometer "reduced to the freezing point."

Collecting the expressions for the several factors above developed, we obtain:

$$R = \alpha_1 \frac{B}{B_0} \frac{\varepsilon^{-1} + \tau_0}{\varepsilon^{-1} + \tau} \tan z \left\{1 - \left[\frac{\alpha_3}{\alpha_1} + \varepsilon h (\tau - \tau_0)\right] \tan^2 z\right\}$$

[1] Chauvenet, Vol. II, p. 165.

From the Pulkowa Tables we find:

$B_0 = 751.5\ mm. \qquad \tau_0 = 9.31^\circ C. \qquad \varepsilon^{-1} = 271.05^\circ C.$

Denoting the quantity enclosed in brackets by F and i troducing numerical values, we obtain:

$$R = \left[1.33207\right] \frac{B F}{271{,}05 + \tau} \tan z$$
$$\log F = -(46.2 + 0.22\,\tau) \tan^2 z \qquad ($$

In the use of these formulæ B and τ must be express in millimeters and degrees C. The formula gives *log F* units of the fifth decimal place. The number enclosed brackets is a logarithm.

The corresponding formulæ, when the pressures are e pressed in English inches and the temperatures in degre F., are:

$$R = \left[2.99215\right] \frac{B F}{455.9 + \tau} \tan z$$
$$\log F = -(42.3 + 0.12\,\tau) \tan^2 z \qquad ($$

The computation by these formulæ is not more laborio than the direct computation from the tables, and the fc lowing comparison shows that the differences between t formulæ and the tables are far less than the uncertainty the tabular numbers themselves. For zenith distances n much exceeding 75° the formulæ may be considered f most purposes a complete equivalent for the tables:

COMPARISON OF THE REFRACTIONS FURNISHED BY THE FORMULÆ AND BY THE PULKOWA TABLES.

Barometer.....	765.0 *mm*	28.500 *in*	765.0 *mm*	28.500 *in*
Att. Thermom.	0.0 C	70.0 F	0.0 C	70.0 F
Ext. Thermom.	– 25.0 C	75.0 F	– 25.0 C	75.0 F
z.........	75°	75°	60°	60°
	″	″	″	″
Tabular Ref...	246.02	192.83	115.36	90.65
Formula A....	246.03	192.84	115.36	90.66
Formula B....	246.02	192.84	115.35	90.66

The coefficients in equations A and B have been so determined as to reproduce with all possible fidelity the refractions of the Pulkowa Tables, but they may be made to represent the actual refractions with greater precision by the application to the constant coefficients of the formulæ of certain corrections depending upon the latitude of the place at which the refraction is required, the amount of moisture in the air and the wave length of the light whose refraction is to be computed. These corrections are developed in Vol. IX, Publications of the Washburn Observatory. The most important of them, and the only one which need be considered here, is that depending upon the latitude. Its effect will be sufficiently taken into account by adding to the bracketed coefficient in the equations A and B, the quantity

$$C = 225 \sin(\varphi - 60^\circ) \sin(\varphi + 60^\circ)$$

where φ denotes the latitude and C is given in units of the fifth decimal place.

II.—TO CORRECT THE SUN'S DECLINATION FOR THE EFFECT OF REFRACTION.

A useful application of the formulæ of the preceding section occurs in connection with the use of the solar compass. It is here required to set off upon a certain divided arc the apparent declination of the sum, *i. e.* the true declination corrected for the effect of refraction. This correction is usually interpolated from rather cumbrous tables of double entry.[1]

Denoting the refraction in declination by d and representing by q the parallactic angle of the sun, we have:

$$d = R \cos q = \left[2.99215\right] \frac{B\,F}{456 + \tau} \tan z \cos q \qquad (1)$$

By applying the fundamental formulæ of spherical trigonometry to the spherical triangle, Pole—Zenith—Sun, and differentiating the equations, we find:

$$\frac{dA}{dt} = \cos \delta \cos q \operatorname{cosec} z \qquad (2)$$

Eliminating $\cos q$ between these equations, we obtain

$$d = \left[2.99215\right] \frac{B\,F}{456 + \tau} \sec \delta \sin z \tan z \frac{dA}{dt} \qquad (3)$$

where z, A, δ and t represent respectively the zenith distance, azimuth, declination, and hour angle of the sun.

The numerical value of $\frac{dA}{dt}$ varies with the position of the sun in the heavens, but may be readily determined at any time as follows: Let the horizontal circle of the solar compass or transit be set to read some integral 10′ and the telescope be than pointed upon the sun by rotating the instrument about the lower motion. The sun having been brought into the field of view, the earth's diurnal motion

[1] See Johnson's Theory and Practice of Surveying, pp. 47, 48.

will carry the sun across the vertical thread of the instrument, and the time at which one edge of the sun is just tangent to the thread should be noted to the nearest second upon a watch. Let the instrument be now turned upon the upper motion, keeping the lower motion clamped, in the direction of the sun's movement, and the vernier set at the next integral 10′. The time at which the sun's edge again becomes tangent to the vertical thread should be noted as before. If we represent by n the interval, in seconds, between the two observed times, we shall have:

$$\frac{dA}{dt} = \frac{40}{n}$$

If desired, the transit may be set so that the second vernier reading is 20′, 30′, etc., greater than the first reading, and we shall then have:

$$\frac{dA}{dt} = \frac{80}{n_2} = \frac{120}{n_3} \ldots. etc. \qquad \text{and}$$

$$n = \tfrac{1}{2} n_2 = \tfrac{1}{3} n_3 \ldots. etc.$$

This value of the differential coefficient enables us to express equation (3) in a form adapted to field use, but since for this purpose an error of even several seconds in the value of d is of small consequence, we shall introduce some modifications in the formula which will render it more convenient without seriously impairing its accuracy. The declination of the sun can never exceed $23\overset{\circ}{.}5$, and we therefore write in the place of $sec\ \delta$ its mean value, 1.051. We also put in place of the temperature τ a mean value, 50° F., and assume for the barometric pressure 30 inches of mercury. With these modifications equation (3) becomes:

$$d = \frac{[3.3854] F \sin z \tan z}{n}$$

We may put the numerator of this fraction equal to *100 N* and tabulate the values of N with the argument the sun's altitude, $h = 90° - z$, as follows:

h	N		h	N	
10°	131′		30°	36′	
		45			14
15	86		40	22	
		24			9
20	62		50	13	
		15			6
25	47		60	7	
		11			4
30	36		70	3	

We now have for the refraction in declination:

$$d = 100 \frac{N}{n}$$

The altitude of the sun, h, should be noted on the vertical circle of the instrument to the nearest half degree at the time of determining n.

The tabulated values of N correspond to a temperature of 50° F. and a barometric pressure of 30 inches. They may be adapted to any other temperature by diminishing d by one per cent for each 5° by which the temperature exceeds 50°, or by increasing one per cent for each 5° below 50°, but this correction and the correction for variation of the barometer can usually be neglected. At great elevations the barometric pressure becomes so much reduced that its variation must be taken account of, and this may be done by diminishing d by one per cent for each 300 feet of elevation above the sea.

The following examples will serve to illustrate the application of the formulæ above developed. On the afternoon of May 12, 1894, at a place in latitude 43° 5′ N., longitude approximately 90° west of Greenwich, I took the following observations with an engineer's transit:

Vernier.		Watch.			Vernier.		Watch.		
°	′	h.	m.	s.	°	′	h.	m.	s.
170	0	4	5	30	170	10	4	23	18
170	10	4	6	27	170	20	4	24	17
Vertical Circle $= 32° 8'$					Vertical Circle $= 28° 46'$				
$n = 57$					$n = 59$				
$N = 33$					$N = 39$				
$d = 58''$					$d = 66''$				

By a direct computation from the formula[1]

$$d = 57' \cot (\delta + N)$$

where N denotes the Bessel auxiliary, I find for the refraction in declination at the time of these observations 59″ and 67″ respectively, thus showing an agreement far within the limits of error permissible in surveying practice.

If, as is often the case, an accuracy of 20″ is sufficient, and the altitude of the sun is not less than 10°, we may dispense with the tabular values of *N* and write

$$d = 2000 \div hn$$

where *h* is the altitude in degrees and the value of *d* is given in minutes of arc. The error of this formula in the preceding cases is 7″ and 4″, respectively.

[1] Chauvenet, Spherical and Practical Astronomy, Vol. I, p. 171.

III.—DETERMINATION OF THE ANGULAR EQUIVALENT OF ONE DIVISION OF A SPIRIT LEVEL.

The methods most in use in this country for the determination of the value of one division of a level require that the level should be attached either to a level-trier or to a telescope provided with a good micrometer. In field astronomy it frequently happens that neither of these auxiliaries is available and the following method, which in respect of precision is not inferior to either of the others, may be employed with advantage since it requires no auxiliary apparatus other than a theodolite or engineer's transit. The original suggestion of this method is supposed to be due to Braun.[1]

Let the spirit level be firmly attached to a theodolite which is thrown out of a level so that its vertical axis makes an angle of from 1° to 3° with the true vertical. It is practically convenient to so attach the level that the radius of curvature drawn through the middle point of its scale shall be approximately parallel to the vertical axis of the theodolite, *i. e.* the level shall be in adjustment. As the theodolite is turned about its vertical axis the level bubble will run from one end of its tube to the other and back again during a complete revolution of the instrument, and two positions, two readings of the azimuth circle, may be found in which the bubble will stand near the middle of its scale. A small turning of the instrument either way from one of these positions will produce a corresponding small motion of the bubble in its tube, and this turning of the theodolite and resulting motion of the bubble may be made to furnish not only the value of a division of the level, but also a test of the uniformity of its curvature.

To determine the relation between the readings of the

---

[1] Astronomische Nachrichten, No. 2490.

azimuth circle of the theodolite and the readings of the level bubble upon its scale, let the accompanying figure

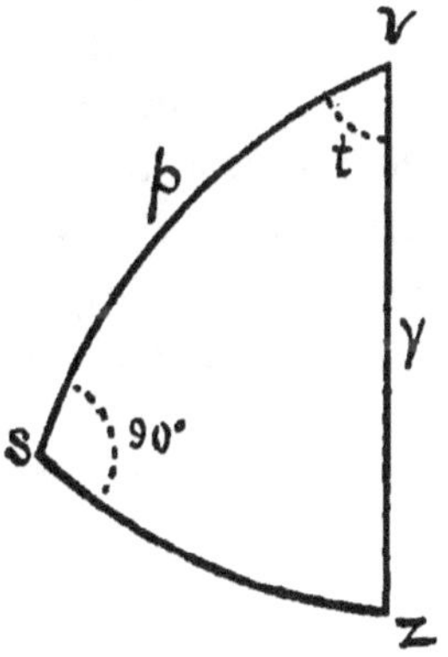

represent a portion of the celestial sphere adjacent to the zenith, Z, and let V and S be the points in which the axis of the theodolite, and the line drawn from the center of curvature of the level tube through the middle of the bubble, respectively, intersect the sphere. The arc SV is the intersection with the celestial sphere of a plane passing through S, V, and the center of curvature of the level tube, and if the adjustment of the level above referred to is approximately made, VS may be considered as the intersection with the sphere of the plane in which the curvature of the level tube lies, so that as the bubble moves in its tube its successive positions, when projected upon the sphere will lie along VS, and any position may be identified by its distance from V, represented in the figure by p. Since the bubble always stands at the highest part of the tube, its position, S, and the corresponding value of p are found by letting fall a perpendicular from the zenith upon the arc VS, and in the right-angled spherical triangle thus formed we have the relation,

$$\tan p = \tan \gamma \cos t$$

where γ, as it appears from the figure, is the angle by which the axis of the theodolite is deflected from the true vertical.

Since the level tube turns with the theodolite when the latter is revolved in azimuth, while the positions of the points V and Z remain unchanged, it appears that the angle t must vary directly with the readings of the azimuth circle, since it measures the inclination of the plane of the level tube to a fixed plane passing through the vertical axis of the instrument. If we represent by A_0 the reading of the circle when the arc VS is made to coincide with VZ, we shall have corresponding to any other reading A′:

$$\tan p = \tan \gamma \cos (A_0 - A') \qquad (1)$$

The value of A_0 in any given case may be determined by finding two positions of the instrument, circle readings A_1 and A_2, in which the bubble stands at the same part of the tube. Since the values of p corresponding to these two readings are equal, we must have:

$$A_0 - A_1 = A_2 - A_0 \quad \text{and} \quad A_0 = \tfrac{1}{2}(A_1 + A_2)$$

If A' and A'' denote slightly different readings of the azimuth circle, and b' and b'' the corresponding readings of the middle of the bubble on the level scale, we may write two equations similar to equation (1), and taking their difference obtain:

$$\frac{\sin (p' - p'')}{\cos p' \cos p''} = 2 \sin \frac{A' - A''}{2} \sin \left(A_0 - \frac{A' + A''}{2} \right) \tan \gamma \qquad (2)$$

Since $p'-p''$ is the distance moved over by the bubble, we may write $p'-p''=(b'-b'')\,d$, where d is the value of a division of the level, and transform (2) into

$$d = \frac{2 \tan \gamma \cos^2 p \sin \frac{1}{2}(A' - A'')}{\sin 1''} \cdot \frac{\sin \left[A_0 - \frac{1}{2}(A' + A'') \right]}{b' - b''} \qquad (3)$$

In this equation $\cos^2 p$ may usually be put equal to 1, or its actual value may be found from the average value of p given by equation (1). Every other factor in the second member of this equation is known with exception of $\tan \gamma$, and the determination of γ will determine d.

For this purpose the instrument should be carefully levelled at the beginning of the work and the telescope directed at some object, approximately at right angles to the line joining two of the leveling screws of the instrument. Let the zenith distance, z', of this object be determined from readings of the vertical circle taken Circle Right and Circle Left. The vertical axis is now to be deflected toward the object by turning the leveling screws, and the zenith distance of the object, reckoned from the vertical axis of the instrument, z'', is to be determined from circle readings in the same manner as z'. We then have, obviously,

$$\gamma = z' - z''$$

To make sure that the deflection of the axis lies in the plane passing through the object sighted upon, it is well to note the position of the bubble of that level of the instrument which is at right angles to the telescope tube. The leveling screws must be so turned that the reading of the bubble of this level on its scale is approximately the same after deflection as before.

By comparison with micrometric apparatus, this determination of γ and the resulting value of d may seem crude, but with a vertical circle reading to minutes only, the values of z' and z'' can be determined within $30''$, and if γ be made 3°, d will be determined with a probable error of one part in four hundred, an accuracy quite sufficient for even the most delicate level. The value of γ should be between 1° and 3°, a coarse vertical circle and fine horizontal circle corresponding to the larger limit, and the reverse conditions to the smaller one.

To illustrate the method, I select the following partial investigation of the microscope level of a small universal instrument, Bamberg No. 2598. The level was investigated by means of the circles of the instrument to which it was attached, without removing or in any way disturbing it:

DETERMINATION OF γ.

Instrument.	Circle R.	Circle L.	z.
Levelled.......	180° 26′ 49″	358° 4′ 13″	91° 11′ 18″
Deflected.......	179 27 3	359 3 48	90 11 37.5

$$\gamma = 0° 59' 40.5''$$

After the level readings which follow were completed, these circle readings were repeated with the instrument deflected and subsequently leveled, giving a second determination of $\gamma = 0° 59' 42''$. I adopt:

$$\gamma = 0° 59' 41''$$

The following are the bubble observations in the deflected position of the instrument:

Bubble.		Circle.	Bubble.		Bubble.		Circle.	Bubble.	
25.3	−0.8	111° 34′	−0.7	25.6	26.3	0.2	291° 6′	0.4	26.0
27.7	1.6	24	1.8	28.1	28.1	2.3	16	1.9	27.7
30.2	4.1	14	3.9	30.1	30.2	4.2	26	3.9	29.7
32.6	6.5	111 4	6.4	32.6	32.3	6.4	36	6.2	32.1
34.8	8.6	110 54	8.7	35.0	34.0	8.1	46	8.5	34.2
37.0	10.9	44	10.8	37.0	36.2	10.2	56	10.6	36.5

The observations began with the level bubble at one end of its scale, circle reading 111° 34′, and the instrument was turned through successive intervals of 10′ until the bubble reached the opposite end, when the settings were repeated in the inverse order to eliminate the effect of any slight change in the instrument or level. The instrument was then turned into the position corresponding to the second set of circle readings which were taken with the bubble running from one end of the tube to the other, in both directions.

The mean of the four readings of the ends of the bubble corresponding to any circle reading may be adopted as the corresponding reading of the middle of the bubble, and these mean readings are given in the following table:

CIRCLE.	BUBBLE.	CIRCLE.	BUBBLE.	τ	$2b$	DIFF.
111°34′	12.35	291° 6′	13.22	89°46′	25.57	
						4.23
24	14.80	16	15.00	89 56	29.80	
						4.28
14	17.08	26	17.00	90 6	34.08	
						4.72
111 4	19.55	36	19.25	90 16	38.80	
						4.18
110 54	21.78	46	21.20	90 26	42.98	
						4.32
44	23.92	56	23.38	90 36	47.30	

Since the bubble readings which stand on the same line in the second and fourth columns of the table are approximately equal, it is apparent that the corresponding circle readings lie on opposite sides of A_0 and equally distant from it. A_0 may, therefore, be determined by taking the mean of any pair of circle readings which stand in the same line, and the angles $A_0—A'$, $A''—A_0$, which we shall designate by τ, may be found by taking half the difference of corresponding circle readings. Values of τ are given in the fifth column of the table.

The quantities $2b$ are the sums of the numbers in the second and fourth columns, and their differences given in the last column show that any irregularities which may exist in the curvature of the level tube are very small, and we may determine a mean value of d to be used over the whole extent of the level tube. Since the values of τ differ so little from 90°, we may assume in equation (3)

$$cos^2 p = 1 \qquad sin\left[A_0 - \tfrac{1}{2}(A' + A'')\right] = 1$$

and taking the differences between the first and fourth, second and fifth, third and sixth lines of the table, we shall have $A'—A''$ constantly equal to 30′, and equation (3) becomes

$$d = \frac{4 \; tan \; \gamma \; sin \; 15'}{2 \, (b' - b'') \; sin \; 1''} = \frac{[1.7959]}{2b' - 2b}$$

from which we obtain the following three values:

$$d = 4.72 = 4.74 = 4.73$$

the mean of which may be adopted.

IV.—THE SIMULTANEOUS DETERMINATION OF FLEXURE, INEQUALITY OF PIVOTS, AND VALUE OF A LEVEL DIVISION FOR A "BROKEN" TRANSIT.

In a "broken" transit, *i. e.* one in which the rays of light are bent at right angles by a reflecting prism placed in the axis, it is well known that the bending of the axis under the weight which it has to carry produces an effect upon the observed times of transit of a star, which may be represented by the expression $f.\cos z \sec \delta$, where f is a constant peculiar to each instrument, and z and δ denote the zenith distance and declination of the star. Since this expression has the same algebraic form as the corrections for inclination of the axis, and for inequality of pivots, they may all be united into a single term:

$$(b' + i + f) \cos z \sec \delta$$

where $\pm(i+f)$ is a constant correction which must be applied to the value of b' directly determined with the spirit level. If $i+f$ is positive for Ocular West it will be negative for Ocular East, and the sign $\pm$ is, therefore, prefixed to it. Since it is not necessary in the use of a broken transit to separate the constant correction $i+f$ into its constituent parts, it will for the present be treated as a single unknown quantity whose value β is to be determined in connection with τ, the angular value of a half division of the level used for measuring b. In a straight transit f is zero, but i has usually an appreciable value and the correction β must, therefore, be determined, and may be conveniently determined by the method here developed for a broken transit.

If from the general equation of the transit instrument[1]

$$\sin c + \sin \delta \sin n - \cos \delta \cos n \sin (\tau - m) = 0 \qquad (1)$$

[1] Chauvenet, Spherical and Practical Astronomy, Vol. II, §123.

the quantities m and n be eliminated by means of the relations (78),[1] we have the following:

$$\sin c + \cos z \sin b - \sin z \cos b \sin (a + A) = 0 \tag{2}$$

where $90^0 - a$ and b represent the azimuth and altitude of the point in which the rotation axis of the instrument, produced toward the west, intersects the celestial sphere. A and z are the azimuth (reckoned from the north toward east) and zenith distance of a star at the instant of its transit over a thread whose collimation is c, *i. e.* the point $90^0 - a$, b is the pole of the small circle traced upon the celestial sphere by the thread in question when the instrument is rotated about its axis, and the distance of this circle from its pole equals $90^0 + c$.

Since in practice b and c are never so great as 10′, equation (2) may be written without sensible loss of accuracy:

$$c + \cos z . b = (a + A) \sin z \tag{3}$$

Substituting in this equation for b its value as given by the spirit level, and writing a similar equation for the case in which the object observed is not the star, but its image reflected from mercury or some other level surface, we have:

$$\begin{aligned} &\textit{Dir.} \quad c' + \cos z' (n' r + \beta) = (a + A') \sin z' \\ &\textit{Ref.} \quad c'' - \cos z'' (n'' r + \beta) = (a + A'') \sin z'' \end{aligned} \tag{4}$$

where n′ and n'' are the measured inclinations of the axis expressed in half divisions of the level scale. We now put

$$z' = z + x \qquad z'' = z - x$$

and introducing these values into(4) find by substraction:

$$\begin{aligned} c' - c'' + (n' + n'') \cos x \cos z . r + 2 \cos x \cos z \beta \\ = (A' - A'') \cos x \sin z + (2a + A' + A'') \sin x \cos z \end{aligned} \tag{5}$$

In practice the object observed will usually be a circumpolar star, and owing to its slow motion the quantity $x = \frac{1}{2} (z' - z'')$ will be so small that we may assume

$$\cos x = 1 \qquad \sin x = \cos \delta \sin t \sin \tfrac{1}{2} (T' - T'')$$

where T' and T'' are the observed times and t is the hour angle of the star at the instant $\frac{1}{2} (T' + T'')$.

[1] *Loc. cit.*

For the coefficient of the last term in equation (*5*) we obtain from (*4*) with sufficient precision

$$2a + A' + A'' = (c' + c'') \, cosec \, z$$

and introducing these values into (*5*) we have

$$(n' + n'') \tau + 2\beta = (A' - A'') \tan z - (c' - c'') \sec z$$
$$+ (c' + c'') \cos \delta \sin t \, cosec \, z \sin \tfrac{1}{2} (T' - T'') \qquad (6)$$

If the star is near the meridian or is observed near the collimation axis of the instrument, the last term in this expression will be very small and may frequently be neglected. Putting

$$P = (A' - A'') \tan z$$
$$Q = (c' + c'') \cos \delta \sin t \, cosec \, z \sin \tfrac{1}{2} (T' - T'')$$

we obtain from the equations

$$\sin z \sin A = - \cos \delta \sin t$$
$$\sin z \cos A = \cos \varphi \sin \delta - \sin \varphi \cos \delta \cos t \qquad (7)$$

reduced by means of the relations furnished by the astronomical triangle, the equation

$$P = \cos \delta \cos q \sec z \, . \, 2 \sin \tfrac{1}{2} (T' - T'') \, 206265$$

where q is the parallactic angle of the star. Introducing Bessel's auxiliary N into this equation, substituting in the last term of (*6*) in place of $\cos \delta \sin t \, cosec \, z$ its equivalent, $\sin A$, and collecting in a form convenient for computation the equations necessary for the reduction of a series of observations, we have the following:

$$\tan N = \cot \varphi \cos t$$
$$P = \left[5.61546 \right] \cos \delta \frac{\sin \tfrac{1}{2} (T' - T'')}{\sin z \tan (N + \delta)} \qquad (8)$$
$$Q = (c' + c'') \sin A \, . \, \sin \tfrac{1}{2} (T' - T'')$$
$$(n' + n'') \tau + 2\beta = P + Q - (c' - c'') \sec z$$

The zenith distance and azimuth of the star, z and A of the formulæ, may either be derived from the instrument at the time of observation, or may be computed from the latitude and the co-ordinates of the star, φ, δ, t, by means of equations (*7*).

Since β changes sign when the instrument is reversed, a similar pair of observations in the reversed position will furnish the equation

$$(n' + n'') \tau - 2\beta = P + Q - (c' - c'') \sec z$$

which, with the last of equations (*8*), suffices for the determination of τ and β. A large change in the inclination of the axis, *e. g.* one which will give values of n' and n'' with altered sign, may be employed for the same purpose. If the inclination of the wyes of the instrument is not disturbed by the reversal, the level readings will furnish directly a determination of the inequality of pivots, and we shall have for the flexure

$$f = \beta - i$$

Formulæ (*8*) become somewhat simplified when the star observed is very near the meridian, but this advantage will often be outweighed by the convenience of observing Polaris at any part of its diurnal path.

The application of the formulæ is illustrated by the following observations of transits of Polaris over the micrometer thread of a large "broken" transit. Each observed time and corresponding micrometer reading is the mean of from five to seven observations made in quick succession. Owing to disturbance of the mercury surface by wind, the reflection observations were difficult and rather discordant. Since the readings of the micrometer diminish in the direction of motion of a star at upper collimation for Ocular West, the collimation corresponding to any reading, R, of the screw is given by the expression

$$c = \pm 57.57 (R - 15) \quad \begin{array}{l} + \textit{Ocular W.} \\ - \textit{Ocular E.} \end{array}$$

The reading of the screw when the thread is in the collimation axis is assumed to be 15.000 *rev.*

WASHBURN OBSERVATORY, OCTOBER 16, 1894.

POLARIS FOR FLEXURE, INEQUALITY OF PIVOTS, ETC.

α = 1 h. 20 m. 55.1 s. φ = 43° 4′ 38″

δ = 88° 44′ 53″.5 *Chronometer* $\triangle T$ = + 3.9s.

log [5.61546] *cos* δ = 3.95484

OCULAR.	WEST.	WEST.	EAST.	EAST.
	h. m. s	*h. m. s.*	*h. m. s.*	*h. m. s.*
T'	20 46 20.7	21 5 52.0	21 32 8.2	22 3 54.0
R'	11.172	8.090	15.274	15.720
n' n''	+ 39.4 + 38.3	− 42.9 − 42.7	− 48.9 − 49.4	+ 40.7 + 40.4
T''	20 35 57.4	21 14 16.7	21 40 39.3	21 53 50.7
R''	12.800	6.289	17.431	13.060
t	19 20 17.8	19 49 13.2	20 15 32.6	20 38 1.1
z	46 30 15	46 21 35	46 14 0	46 8 5
A	1 37	1 32	1 26	1 20
log cos t	9.53558	9.66155	9.74628	9.80355
N	20 9 25	26 7 56	30 48 25	34 13 43
log cosec z	0.13941	0.14045	0.14136	0.14208
log cot $(N + \delta)$	9.53463_n	9.66630_n	9.75362_n	9.81213_n
log sin $\frac{1}{2}(T' - T'')$	8.35530	8.26364_n	8.26911_n	8.34113
log sin A	8.452	8.429	8.400	8.369
log $(c' + c'')$	2.541_n	2.955_n	2.193_n	1.848
log P	1.98618_n	2.02523	2.11893	2.25018_n
log Q	9.348_n	9.648	8.862	8.558
log $(c' - c'')$	1.97185_n	2.01571	2.09405	2.18508_n
log sec z	0.16222	0.16107	0.16007	0.15929
	″	″	″	″
$(c' - c'')$ *sec z*	− 136.17	+ 150.24	+ 179.52	− 220.99
P	− 96.87	+ 105.98	+ 131.50	− 177.90
Q	− 0.22	+ 0.44	+ 0.07	+ 0.04

The preceding computation furnishes the absolute terms of the following equations:

$$\begin{aligned} +77.7\,\tau + 2\beta &= +39^{s}.08 \\ -85.6\,\tau + 2\beta &= -43.82 \\ -98.3\,\tau - 2\beta &= -47.95 \\ +81.1\,\tau - 2\beta &= +43.13 \end{aligned}$$

A least square solution of these equations furnishes the values:

$$\tau = +0^{s}.506 \qquad \beta = -0^{s}.600$$

From numerous determinations with the spirit level, the inequality of the pivots is known to be $i = -0''.64$, which, combined with the value of β, gives for the flexure the value $f = +0''.04$.

V.—DETERMINATION OF TIME AND AZIMUTH FROM TRANSITS OVER THE VERTICAL OF THE POLE STAR.

In a development of the formulæ for determining the time from transits over the vertical of a circum-polar star, published in 1828, Bessel says by way of introduction: "That this may not appear futile I remark, what Hansteen and Schumacher have properly noted, that the most appropriate use of a portable transit instrument for a time determination consists in mounting it, not in the meridian, but in an azimuth which admits of an observation of one of the polar stars, wherever this may be with respect to the meridian, closely followed or preceded by a transit of a fundamental star."

The obvious advantage which this mode of observing possesses lies in the shorter period of time during which the observer depends upon the stability of his instrumental constants. For meridian observations this period is rarely much less than half an hour, while by the method suggested it need never exceed five minutes. Nevertheless, the general opinion of two generations of field astronomers seems fairly represented by the words of Chauvenet, who, after devoting a score of pages to a discussion of the method, remarks in closing: "The methods which have here been given * * * are intended for the use of observers in the field who have but little time to adjust their instruments and wish to collect all the data possible, reserving their reduction for a future time. The greater labor of these reductions, compared with those of meridian observations, is often more than compensated by the saving of time in the field." This greater labor of reduction is now obviated through the simplifications introduced into the method by the Russian astronomer, Döllen, who maintains with equal zeal and cogency the greater precision and at least equal convenience of his method for all purposes of field astronomy. Under Döllen's influence the method has, within the last quarter century, come into considerable use in eastern and central Europe, and from an extended practical application of it the writer of these pages is satisfied of the justice of the claims made in its behalf. This section of the present paper is an attempt to bring to the attention of American teachers of practical astronomy, in substance, the theory of Döllen's method, but it cannot be considered a substitute for the precepts and discussion contained in the elaborate introduction to the *Stern Ephemeriden zur Bestimmung von Zeit und Azimut*, published annually by Döllen since 1886.

As indicated by the above title, the observations for time are equally available for a determination of azimuth, and reduced to their simplest terms these observations are as follows: Let the transit (universal instrument, or theodolite, in case a determination of azimuth is also desired) be pointed at Polaris, and the chronometer time, S', at which the star appears bisected by the middle vertical thread, noted. Then revolve the telescope about the horizontal axis without disturbing the azimuth of the instrument and observe the time of transit, S, of a clock star over all of the threads, and measure the inclination of the axis, b, with a spirit level, if possible both before the observation of Polaris and after that of the southern star. Reverse the instrument, point again upon Polaris, and observe it and a clock star, as before. If the instrument possess a graduated horizontal circle, which is read in connection with the observations of the stars, these data will determine the zero point of the circle, *i. e.* its reading when the telescope points north, and the azimuth of any terrestrial point toward which the telescope may be directed.

We proceed to consider the theory of the method and adopt as a basis for the investigation the fundamental equation of the transit instrument,[1]

[1] Chauvenet, Vol. II, Eq. (79).

$$\sin(\tau - m) = \tan n \tan \delta + \sin c \sec n \sec \delta \qquad (1)$$

together with the equations

$$\tan n = \sin b \sec n \operatorname{cosec} \varphi - \sin m \cot \varphi \qquad (2)$$

$$\cos a \tan m = \tan b \cos \varphi + \sin \varphi \sin a \qquad (3)$$

furnished by the spherical triangle, PZA, formed by the pole, the zenith and the point in which the rotation axis of the instrument, produced toward the west, intersects the celestial sphere. The sides and angles of this triangle have the following values:

$$PZ = 90^\circ - \varphi \qquad PA = 90^\circ - n \qquad ZA = 90^\circ - b$$

$$P = 90^\circ - m \qquad Z = 90^\circ + a$$

The symbol τ represents the east hour angle of the star at the instant of transit over the middle thread, and we have obviously the relation

$$\tau = \alpha - S - \Delta T \qquad (4)$$

Since each star observed furnishes an equation of the types (*1*) and (*4*), it appears that if the instrumental constants b and c are known an observation of the transits of a circum-polar star and a southern star suffice for the determination of the unknown quantities ΔT, m, n, a, and our problem consists solely in so transforming the preceding equations as to facilitate the determination of ΔT and a.

Denoting by the subscripts 1 and 2, respectively, quantities pertaining to the polar and the southern star, we write equation (*1*) for each of these stars as follows:

$$\left.\begin{aligned} \sin(\tau_1 - m - \vartheta) &= \tan \delta_1 \tan n \left\{ 1 + \operatorname{cosec} \delta_1 \operatorname{cosec} n \sin(c + x_1) \right\} \\ \sin(\tau_2 - m - \vartheta) &= \tan \delta_2 \tan n \left\{ 1 + \operatorname{cosec} \delta_2 \operatorname{cosec} n \sin(c + x_2) \right\} \end{aligned}\right\} \quad (5)$$

where ϑ, x_1, and x_2 are small arbitrary quantities subject only to the condition that they must be so determined as to satisfy the equations. Since this is equivalent to only two relations among the three quantities we are at liberty to impose a third relation, for which we choose

$$\sin(c + x_1) \sin \delta_2 = \sin(c + x_2) \sin \delta_1$$

which makes the bracketed factors in the two equations equal. Presupposing that ϑ, x_1, and x_2 are small quantities we differentiate equations (*5*), and eliminating x_1, and x_2 find, when quantities of the order cn^2 are neglected,

$$\vartheta = \frac{(1 - \sin\delta_2)\,c}{\cos\delta_2 - \sin\delta_2 \cot\delta_1 \cos(\tau_1 - m)}$$

If for δ_2 we substitute the polar distance, $p_2 = 90^\circ - \delta_2$, this equation becomes, very approximately,

$$\vartheta = c \, . \tan\tfrac{1}{2}p_2 \left\{1 + \cot\delta_1 \tan\delta_2 \cos(\tau_1 - m)\right\} \qquad (6)$$

Dividing the first of equations (*5*) by the second, we obtain:

$$\tan\left[\tfrac{1}{2}(\tau_1 + \tau_2) - m - \vartheta\right] = \frac{\sin(\delta_1 + \delta_2)}{\sin(\delta_1 - \delta_2)} \tan\tfrac{1}{2}(\tau_1 - \tau_2) \qquad (7)$$

We now assume the auxiliary quantities,

$$2\tau = (\alpha_1 - S') - (\alpha_2 - S)$$
$$U = \alpha_2 - S - \varDelta T - m - \vartheta \qquad (8)$$

and introducing them into (*7*) find

$$\tan(\tau + U) = \frac{\sin(\delta_1 + \delta_2)}{\sin(\delta_1 - \delta_2)} \tan\tau$$

whose solution is

$$\tan U = \frac{\cot\delta_1 \tan\delta_2 \sin 2\tau}{1 - \cot\delta_1 \tan\delta_2 \cos 2\tau} \qquad (9)$$

In equations (*8*) $\varDelta T + m$ is now the only unknown quantity, and to determine m we apply (*1*) to the polar star and substitute in it the value of $\tan n$ given by (*2*) and the value of $\tau_1 - m$ given by (*4*) and (*8*), and find

$$\sin m = -\cot\delta_1 \tan\varphi \sin(2\tau + U + \vartheta) + \sin b \sec\varphi + \sin c \tan\varphi$$

in which terms of the order cn^2 are neglected. Subtracting from each member of the equation the auxiliary quantity

$$\sin m' = -\cot\delta_1 \tan\varphi \sin(2\tau + U) \qquad (10).$$

we obtain to the same degree of approximation

$$m = m' + b\sec\varphi + c\tan\varphi - \vartheta\cot\delta_1 \tan\varphi \cos(2\tau + U)$$

Substituting for ϑ its value in terms of c, and introducing into (*8*) the resulting value of m, we obtain

$$\varDelta T + Cc = \alpha_2 - (S + U + m' + b\sec\varphi) \qquad (11)$$

where the coefficient C has the value

$$C = \tan\varphi + \tan\tfrac{1}{2}p_2 \left\{ 1 + (\tan\delta_2 - \tan\varphi)\cot\delta_1 \cos(2\tau + U) \right\} \quad (12)$$

If at the time of observation the southern star was near the zenith, or Polaris was near elongation, or the collimation constant, c, was very small, the bracketed factor may be put equal to 1, giving

$$C = \tan\varphi + \tan\tfrac{1}{2}p_2$$

For a determination of azimuth we write equation (3) in the form

$$\tan a = \tan m \operatorname{cosec}\varphi - \tan b \cot\varphi$$

and assuming the equation

$$\tan a' = \tan m' \operatorname{cosec}\varphi \quad (13)$$

find by subtraction

$$a = a' + b\tan\varphi + c\sec\varphi \left\{ 1 - \cot\delta_1 \tan\tfrac{1}{2}p_2 \cos(2\tau + U) \right\} \quad (14)$$

If K and M denote respectively the reading of the azimuth circle corresponding to the star observations, and to that position of the instrument in which the rotation axis lies in the plane of the prime vertical (collimation axis in the meridian), we have, obviously,

$$M = K + a' + b\tan\varphi + C'c \quad (15)$$

where C' is an abbreviation for the coefficient of c given in the preceding equation.

Since the collimation constant, c, changes sign when the instrument is reversed, an observation of Polaris and a southern star in each position of the instrument, W. and E., will suffice for the determination of ΔT and c from the observed times of transit, and also, if the instrument is provided with an azimuth circle, for the determination of M and c, from the circle readings. The agreement between the two values of c thus determined furnishes a valuable control upon the accuracy of the observations and their reduction.

In the preceding investigation the effect of flexure, inequality of pivots and diurnal aberration has been neglected. These quantities may, however, be taken into account, as in the case of meridian observations, by applying to the observed level constant, b, a correction, $\pm\beta$, for the first two sources of error, and by applying to S a correction,

$$-\overset{\text{s.}}{0.021} \cos\varphi \,.\, C$$

for the aberration.

The formulæ requisite for the reduction of observations in the vertical of the pole star may now be collected, slightly simplified and arranged as follows:

Data known independently of the observations:

$$\varphi,\ \alpha_1,\ \alpha_2,\ \delta_1,\ \delta_2,\ \varkappa = \overset{\text{s.}}{0.021} \cos\varphi,\ p_2 = 90^\circ - \delta_2$$

Data given by the observations: S', S, b, K.

$$\begin{aligned}
t &= (\alpha_1 - \alpha_2) + (S - S') \\
h &= 1 + \tan\delta_2 \cot\delta_1 \cos t \\
l &= 1 - \tan\tfrac{1}{2}p_2 \cot\delta_1 \cos t \\
C &= h \tan\tfrac{1}{2}p_2 + l \tan\varphi \\
C' &= 15\, l \sec\varphi \\
\tan U &= \frac{\cot\delta_1 \tan\delta_2 \sin t}{1 - \cot\delta_1 \tan\delta_2 \cos t} \qquad (16) \\
-\sin m' &= \tan\varphi \cot\delta_1 \sin(t + U) \\
\tan a' &= \tan m' \operatorname{cosec}\varphi \\
\Delta T + Cc &= \alpha_2 - (S + U + m' + b\sec\varphi - C\varkappa) \\
M - C'c &= K + a' + b\tan\varphi - C'\varkappa
\end{aligned}$$

The computation of these formulæ may be somewhat facilitated by an algebraic device upon which Döllen places great stress. From the ordinary development of $\sin x$ and $\tan x$ in series, we have, when x is small,

$$\log\sin x = \log x - \frac{Mx^2}{6} \qquad \log\tan x = \log x + 2\,\frac{Mx^2}{6}$$

where M denotes the modulus of the common system of logarithms. Putting

$$\sigma = \tfrac{1}{6} Mx^2$$

we may tabulate σ with x or $\log x$ as argument, and such a table is given by Döllen with $\log x$, when x is expressed in seconds of time, as argument. When x is expressed in arc values of σ may be taken from any logarithmic table by means of the relation

$$\sigma = \tfrac{1}{3} (\log \tan x - \log \sin x)$$

If $\sigma(U)$ denote the value of σ corresponding to $\log U$ when U is expressed in seconds of time we may, by the introduction of the divisor, $15 \sin 1''$, obtain in seconds of time and arc, respectively,

$$\begin{aligned} U &= \left[4.13833 - 2\sigma(U)\right] \frac{\cot \delta_1 \tan \delta_2 \sin t}{1 - \cot \delta_1 \tan \delta_2 \cos t} \\ (-m') &= \left[4.13833 + \sigma(m')\right] \cot \delta_1 \tan \varphi \sin (t + U) \qquad (17) \\ \log a' &= \log (15 \operatorname{cosec} \varphi) + \log m' + 2\sigma(m') - 2\sigma(a') \end{aligned}$$

In equations (16) the quantities h, l, C, C' are analogous to the transit factors A, B, C used for the reduction of meridian observations, and C, C' may be tabulated for a given latitude and assumed constant for a period of several years. The quantities U and m' must be computed anew for each observation, and a' must also be computed in case the azimuth is required. To diminish the labor of this computation Döllen tabulates for a selected list of 180 stars certain General Constants, through which these computations are considerably shortened.

With assumed values of the coördinates of the stars and an assumed interval $S - S' = 4^m$ put

$$-U = x_0 \qquad + \frac{206265}{15} \cot \delta_1 \sin (t + U) = N_0$$

We shall then have

$$-(U + m') = x_0 + p N_0 = t_0 \qquad a' = p' N_0$$

where p and p' are functions of the latitude which differ from $\tan \varphi$ and $15 \sec \varphi$ by terms of the kind above represented by σ,

$$\begin{aligned} \log p &= \log \tan \varphi + \sigma (N \tan \varphi) \\ \log p' &= \log (15 \sec \varphi) + 3\sigma (N \tan \varphi) - 2\sigma (N \sec \varphi) \end{aligned}$$

The values of p and p' may be conveniently tabulated for a given latitude with $log\,N$ as the argument, and for this purpose $log\,p'$ is best expressed in the form

$$log\,p' = log\,(15\,sec\,\varphi) + \sigma\,(N\sqrt{tan^2\varphi - 2})$$

where the two σ terms given above have been united into a single term whose numerical value is to be obtained regardless of the sign of the quantity under the radical, and then to be added or subtracted as this quantity is positive or negative. The following is such a table for the latitude of the Washburn Observatory, $\varphi = 43°\,4'\,37''$, and it should also be noted that the values of $log\,N$ are limiting values at which the tabular p, p' changes from one value to the next:

p	$log\,N$	p'	$log\,N$
9.97083		1.31251	
	2.238		1.921
.97084		.31250	
	2.381		2.239
.97085		.31249	
	2.476		2.363
.97086		.31248	
	2.539		2.442
.97087		.31247	
	2.587		2.499
.97088		.31246	

The construction of such a table is the only point at which the σ terms are required in the application of Döllen's ephemerides.

In general the coördinates of the stars and the observed interval $S - S'$ will differ from that assumed in the computation of x_0 and N, and it will be convenient to pass from these latter quantities to the values x, N corresponding to the actual observation by means of differential formulæ.

It is evident from an inspection of equations (*16*) that these differential formulæ will contain some terms which involve only the coördinates of the stars and are, therefore, the same for all parts of the earth's surface, while other terms will involve functions of the latitude, and only that part of these terms which is independent of the latitude can conveniently be tabulated. Leaving the reader to differentiate for himself equations (*16*), we reproduce here the form in which Döllen expresses the differential coefficients and the correction terms involving them:

$$x + pN = t_0 + Qk + RG + D\,\Delta\,\delta$$
$$a' = p'N = p'N_0 + Q'k + R'G$$

where k, G and $\Delta\,\delta$ represent variations in the elements with which x_0 and N_0 were computed, and Q, Q', R, R', D are differential coefficients having the following values:

$$Q = p\lambda + \beta \qquad R = p\mu + \gamma$$
$$Q' = p'\lambda \qquad R' = p'\mu$$

The values of $\beta, \gamma, \lambda, \mu$ and D involve only the coördinates of the stars and are given among the general constants for each star of Döllen's list.

The values of k, G and $\Delta\,\delta$ are as follows:

$$\Delta\,\alpha = \alpha_2 - (\alpha_2)_0 \qquad g = -\left\{\alpha_1 - (\alpha_1)_0\right\}$$
$$\Delta\,\delta = \delta_2 - (\delta_2)_0 \qquad k = -\left\{\delta_1 - (\delta_1)_0\right\} \qquad G = g + \Delta\,\alpha$$

where the subscript $_0$ denotes the tabular values of the coördinates corresponding to x_0, N_0. These assumed values are given as a part of the table of constants for each star, and an ephemeris of g and *log* k precedes the table of constants.

The actual reduction of a set of observations by means of these general constants will not often be made, but recourse will be had to the General Ephemerides constructed from them for 93 of the 180 stars. These ephemerides give at intervals of ten days throughout the year the instantaneous values of N and T, $T = \alpha_2 + x$, and from them the observer should construct a local ephemeris of the values of θ and a' for a few of the tabular dates near the epoch of his observations, using the relations

$$\theta = T + pN \qquad a' = p'N$$

Values of θ and a' interpolated from the local ephemeris will be immediately available for the reduction of observations in which the observed interval $S - S'$ equals the value 4^m assumed in the computation of x_0 and N_0 The observations should be so arranged as to secure at least a rough approximation to this interval between the observation of Polaris and the clock star, but a deviation of even several minutes from the prescribed amount may be very simply corrected.

Since the interval $S - S'$ affects U, m' and a' precisely as does $\alpha_1 - \alpha_2$ whose effect is represented in the term RG, we apply to S and K the corrections

$$R \left\{ S - (S' + 4^m) \right\} \qquad R' \left\{ S - (S' + 4^m) \right.$$

and the reduction of the observations takes the very simple form:

$$r = \frac{1}{100} \left\{ S - (S' + 4^m) \right.$$

$$S_0 = S + R_0 r + Bb - C\varkappa \qquad \Delta T \pm Cc = \theta - S_0$$

$$K_0 = K + R'_0 r + B'b - C'\varkappa \qquad M \mp C'c = K_0 - a'$$

The level corrections Bb, $B'b$ are most conveniently taken from a table of multiples of

$$\frac{\tau}{30} \sec \varphi = B\tau \qquad \frac{\tau}{2} \tan \varphi = B'\tau$$

where τ represents the angular value of one division of the level scale. The factor R_0 equals $100\,R$ and its value together with that of the collimation factors C, C' are to be derived from the data given with each star in the ephemeris

$$R_0 = p\mu_0 + \gamma_0 \qquad C = pC_1 + C_0$$

$$R'_0 = p'\mu_0 \qquad C' = p'C_1$$

These values when once computed should be preserved for future use.

The reduction to the middle thread of transits of a clock star observed over the side threads must not be made, as in the meridian, by the use of the factor C, but by a special factor F whose logarithm is given in the ephemeris and among the general constants for each star.

$$F = \sec \delta_2 \sec n \sec r$$

Certain auxiliary quantities to be used in setting the instrument so as to find the stars to be observed are also given in the tables. Their use will be understood from the following precept: "At the sidereal time $\theta - 4^m$ point upon the pole star by means of its azimuth a' and zenith distance $z' = H - (\varphi + \nu \tan \varphi)$ and without changing the azimuth of the instrument await the clock star at the zenith distance $z = \rho - z'$."

The following two examples illustrate, respectively, the application of the trigonometric formulæ, equations (*16*) and (*17*), and of Döllen's ephemerides, to the reduction of observations made with a very small universal instrument, having an objective with a clear aperture of 35 *mm*, focal length 373 *mm*, magnifying power of ocular 36 diameters, azimuth circle read by estimation to single seconds. In view of the small dimensions and feeble power of the instrument the agreement between the values of the collimation constant c given by the observed times and the circle readings is sufficiently satisfactory.

The computation by the trigonometric formulæ is so arranged that the values of U, m', etc., may be obtained either with or without the use of the σ terms.

1891, SEPTEMBER 4. OBSERVER, G. C. C.

BAMBERG UNIVERSAL INSTRUMENT.

φ = 43° 4′ 47″ α_1 = 1h. 19m. 33s. δ_1 = 88° 43′ 34.7″

$\log\tan\varphi$ = 9.97082 $\log\text{cosec}\,\varphi$ = 0.16559 $\log\cot\delta_1$ = 8.34702

Star. Oc.	*ε Cygni W.*	*ζ Cygni E.*			
α_2	20 41 50.19	21 8 20.00	$\alpha_1 - \alpha_2$	4 37 43	4 11 13
δ_2	+ 33 33 57	+ 29 47 1	$S - S'$	4 28	4 35
C C'	1.47 20.7	1.51 20.7	t	70 32 45	63 57 0
S'	20 38 56	21 5 39	$\cos t$	9.52251	9.64262
S	20 43 23.59	21 10 14.03	$\tan\delta_2$	9.82187	9.75764
	″ s.	″ s.	$\sin t$	9.97447	9.95348
b . $b\sec\varphi$	− 7.2 − 0.65	+ 5.1 + 0.46	$(\cot\delta_1\tan\delta_2\cos t)$	7.69140	7.74738
U	+ 3 12.22	+ 2 38.08	$1-(\ )$	9.99786	9.99757
. m'	− 4 30.87	− 4 18.27	$\cot\delta_1\tan\delta_2\sin t$	8.14336	8.05814
	s. ″	s. ″			
$C\varkappa$ $C'\varkappa$	0.02 0.3	0.02 0.3	$\tan U$	8.14550	8.06057
K	344 38 31.5	154 33 44.0	2σ	3	2
a'	− 1 39 8.1	− 1 34 31.6	$\sin(t+U)$	9.97656	9.95588
$b\tan\varphi$	− 6.7	+ 4.8	$-\sin m'$	8.29440	8.27372
$\Delta T \pm Cc$	− 14.08	− 14.28	σ	3	3
$M \mp C'c$	332 59 16.4	152 59 16.9	$\tan a'$	8.46008	8.43939

$\Delta T + 1.47\,c = -14.08$ s. $M - 20.7\,c = 16.4$″

$\Delta T - 1.51\,c = -14.28$ $M + 20.7\,c = 16.9$″

$\Delta T = -14.18$ s. $M = 16.6$″

$c = +\,0.07$ s. $c = +\,0.01$ s.

1891, September 4.

$\log p = 9.97085$ $\log p' = 1.31247$

Star Oc.	ε Cygni W.	ζ Cygni E.	Equations:
R R'	− 0.196 − 14.85	− 0.350 − 19.85	s.
C C'	1.471 20.46	1.513 20.42	$\Delta T + 1.47\,c = -14.08$
θ	20 43 8.79	21 10 0.06	$T - 1.51\,c = -14.29$
	″	″	
b	− 7.2	+ 5.1	$\Delta T = -14.18$
$S' + 4^m$	20 42 56	21 9 39	$c = +0.07$
S	20 43 23.59	21 10 14.03	$M' - 20.5\,c = 16.3$
$C\kappa.$ $R_0 r$	− 0.02 − 0.05	− 0.02 − 0.12	$M' + 20.4\,c = 16.7$
$b \sec \varphi$	− 0.65	+ 0.46	$M = 16''.5$
			s.
$\Delta T \pm Cc$	− 14.08	− 14.29	$c = +0.01$
	° ′ ″	° ′ ″	° ′ ″
a'	1 39 4.1	1 34 24.8	$M = 332$ 57 16.5
K	334 38 31.5	154 33 44.0	
$C'\kappa$ $R'r$	− 0.3 − 4.1	− 0.3 − 7.0	
$b \tan \varphi$	− 6.7	+ 4.8	
$M \mp C'c$	332 59 16.3	152 59 16.7	

VI.—DETERMINATION OF LATITUDE AND TIME FROM EQUAL ALTITUDES OF STARS.

The simultaneous determination of time and latitude from the observed instants at which three different stars reach the same (unknown) altitude is discussed in the principal text books of spherical astronomy, but the laborious character of the reduction of the observations there developed has prevented the method from coming into general use, although from theoretical considerations and from experience it has been abundantly shown to furnish a very accurate determination of both time and latitude. In the following pages an attempt is made to simplify the method by substituting for the observation of three stars separated by considerable intervals of time the observation of the time at which a single star transits over the almucantar of a close circum-polar star, usually Polaris, the elapsed time between the pointing of the instrument upon the polar star and the observed transits of the clock star being made as short as possible, *e. g.* five minutes, or less.

Such a comparison of one clock star with one polar furnishes a single relation between the latitude and the clock correction, and a similar comparison of another star furnishes a second relation which suffices for the determination of both quantities. It should be noted that these two sets of observations are entirely independent of each other and require no assumption with regard to the stability of the instrumental constants, save for the brief interval between pointing upon Polaris and observing the southern star.

The almucantar and the zenith telescope are the instruments best adapted to observations of this kind, but any instrument which possesses a telescope rotating about a horizontal and a vertical axis and provided with a level whose plane is perpendicular to the horizontal axis, may be used, *e. g.* a universal instrument or an engineer's transit. If the makers would furnish a simple means of fastening the striding level which accompanies the better class of transits, with its tube at right angles to the horizontal axis, the efficiency of these instruments would be very greatly increased, but even without this attachment the observation of equal altitudes is the most advantageous mode of employing such an instrument for the determination of either latitude or time. We proceed to develop the equations for the general case in which both of the quantities are required.

Let T_1 and T_2 denote the observed times at which two stars cross a given almucantar whose (unknown) zenith distance is z, and let $\alpha_1, \pi, \alpha_2, p$, be the right ascensions and polar distances of the northern and southern star, respectively. The formulæ for the transformation of coördinates furnish for the two stars the equations:

$$\begin{aligned} \cos z &= \sin \varphi \cos \pi + \cos \varphi \sin \pi \cos (T + \tau) \\ &= \sin \varphi \cos p + \cos \varphi \sin p \cos (T - \tau) \end{aligned} \qquad (1)$$

where

$$T + \tau = T_1 + \Delta T - \alpha_1 \qquad T - \tau = T_2 + \Delta T - \alpha_2$$

Subtracting the second equation from the first and dividing by

$$2 \sin \tfrac{1}{2} (p + \pi) \sin \tfrac{1}{2} (p - \pi) \cos \varphi$$

we obtain

$$\tan \varphi = \cot \tfrac{1}{2} (p + \pi) \cos T \cos \tau - \cot \tfrac{1}{2} (p - \pi) \sin T \sin \tau \qquad (2)$$

We introduce into this equation the auxiliaries

$$l \cos \lambda = \cot \tfrac{1}{2} (p + \pi) \cos \tau \qquad l \sin \lambda = \cot \tfrac{1}{2} (p - \pi) \sin \tau \qquad (3)$$

and obtain

$$l \cos (T - \lambda) = \tan \varphi \qquad (4)$$

From equations (3) we obtain

$$\begin{aligned} l \sin (\lambda - \tau) &= \left\{ \cot \tfrac{1}{2} (p - \pi) - \cot \tfrac{1}{2} (p + \pi) \right\} \sin \tau \cos \tau \\ l \cos (\lambda - \tau) &= \cot \tfrac{1}{2} (p - \pi) \sin^2 \tau + \cot \tfrac{1}{2} (p - \pi) \cos^2 \tau \end{aligned} \qquad (5)$$

which furnish, after a little reduction,

$$\tan(\lambda - \tau) = \frac{\sin\pi \sin 2\tau}{\sin p - \sin\pi\cos 2\tau} \qquad (6)$$

We also obtain from (3)

$$\left\{\frac{\cos\lambda}{\cos\tau} + \frac{\sin\lambda}{\sin\tau}\right\} l = \frac{\sin p}{\sin\frac{1}{2}(p+\pi)\sin\frac{1}{2}(p-\pi)} \qquad (7)$$

from which

$$l^{-1} = \frac{\cos\pi - \cos p}{\sin p} \cdot \frac{\sin(\tau+\lambda)}{\sin 2\tau} \qquad (8)$$

In this expression we put

$$\frac{\cos\pi - \cos p}{\sin p} = \tan\tfrac{1}{2}(p - x)$$

and find the rigorous equation

$$\tan\tfrac{1}{2}x = \tan^2\tfrac{1}{2}\pi \cot\tfrac{1}{2}p \qquad (9)$$

for which there may usually be substituted

$$x = \frac{2\sin^2\frac{1}{2}\pi}{\sin 1''}\cot\tfrac{1}{2}p$$

Introducing (8) into (4) it becomes

$$\cos(T - \lambda) = \tan\varphi \tan\tfrac{1}{2}(p - x)\frac{\sin(\tau+\lambda)}{\sin 2\tau} \qquad (10)$$

We now put

$$\lambda - \tau = M \qquad T - \lambda = N$$

and obtain

$$T - \tau = T_s + \Delta T - \alpha_s = M + N \qquad (11)$$

These equations suffice for the determination of ΔT when the latitude, φ, is known, and the effect upon ΔT of an error in the assumed value of φ is readily shown to be

$$\frac{d\,\Delta T}{d\,\varphi}\Delta\varphi = \frac{d\,N}{d\,\varphi}\Delta\varphi = -2\,\mathrm{cosec}\,2\,\varphi\cot N \,.\, \Delta\varphi \qquad (12)$$

Putting $t = 2\tau$ and eliminating the formulæ requisite for the reduction of an observation may be collected and arranged as follows:

$$t = (\alpha_2 - T_2) - (\alpha_1 - T_1)$$

$$a = \frac{\sin \pi}{\sin p} \qquad x = \frac{2 \sin^2 \frac{1}{2} \pi}{\sin 1'} \cot \tfrac{1}{2} p$$

$$\tan M = \frac{a \sin t}{1 - a \cos t} \tag{13}$$

$$\cos N = \frac{\tan \varphi \tan \frac{1}{2} (p - x) \cos M}{1 - a \cos t}$$

$$C = 2 \operatorname{cosec} 2 \varphi \cot N$$

$$\Delta T + C \Delta \varphi = (\alpha_2 - T_2) + (M + N)$$

Since $\cos N = \cos (-N)$ the algebraic sign of N is not determined by the equations, but it is apparent from the physical conditions of the problem that N must be positive for a star west of the meridian and negative for a star east of the meridian. The coefficient of $\Delta \varphi$ appears in equations (13) with changed sign in order that $\Delta \varphi$ may represent a correction to the assumed latitude. In the use of these formulæ x may be computed with four place logarithms, a and M with five place, and N with six or seven place tables.

It will frequently happen that the observations of the polar star and the southern star composing a pair will be made at slightly different zenith distances, the slight change in the zenith distance of the line of sight of the telescope being indicated by an altered reading of the level bubble. This alteration is most conveniently taken into account by applying to the observed time, T_2, a correction

$$\text{Level Corr.} = \frac{dt}{dz} \Delta z = \frac{\sec \varphi}{\sin A_2} \cdot \frac{(b_2 - b_1)\tau}{15} \tag{14}$$

where b_2 and b_1 are the level readings, τ the value of a level division in seconds of arc, and A_2 the azimuth of the star. The factor $b_2 - b_1$ is to be considered positive when the bubble runs from its initial position toward the objective end of the telescope.

The factor $\frac{\sec \varphi}{\sin A_2}$ in the preceding equation may be replaced by an expression which is most conveniently treated in connection with the thread intervals.

The southern star, and occasionally the polar star, will be observed on several threads, and from the several observed times the time of transit over the middle thread may be found by Bessel's method,[1] or as follows: The reduction of any thread to the middle thread is given by the equation

$$T_2 = T + i\frac{dT_2}{dz} + \frac{i^2}{2}\frac{d^2T_2}{dz^2} + \ldots.$$

$$= T + i \sec\varphi \operatorname{cosec} A_2 + \frac{i^2}{2}\frac{\sec\varphi \operatorname{cosec} A_2 \cot A_2}{\sin z \tan q} \qquad (15)$$

where q is the parallactic angle of the star when on the middle thread.

When the star is observed at its transit over the almucantar passing through the pole, we have rigorously

$$q = t \qquad z = 90^\circ - \varphi$$

and since the last term of (15) is very small we may in most cases substitute these approximate values in it. From the observations on the first and last threads we obtain, approximately,

$$f = \sec\varphi \operatorname{cosec} A_2 = (T'' - T') \div (i'' - i') \qquad (16)$$

Applying (15) to each observed thread and taking the mean of the resulting equations, we obtain

$$T_2 = \frac{1}{n}\left\{\Sigma T + f \Sigma i\right\} + f^2 \cos A \cot t \frac{1}{n} \Sigma \frac{2 \sin^2 \frac{1}{2} i}{\sin 1''} \qquad (17)$$

The last term rarely amounts to more than a few hundredths of a second, and if the star observed is near the prime vertical, or near elongation, it may be neglected. It should be noted that owing to the factor $\operatorname{cosec} A_2$, f is positive for stars west of the meridian and negative for stars east of the meridian.

Effect of Diurnal Aberration.—The effect of the diurnal aberration is to displace every star toward the east point of the horizon by the amount

$$D = \overset{s.}{0.021} \cos\varphi \sin \Delta$$

[1] Chauvenet, Table VIII.

where Δ is the angular distance of the star from the east point. If in the quadrantal triangle formed by the star, the zenith and the east point we represent the angle at the star by ψ we shall have for the effect of the diurnal aberration upon the time of the star's transit over an almucantar

$$\varkappa = 0\overset{s.}{.}021 \cos\varphi \sin\Delta \,.\, \cos\psi \,.\, \sec\varphi \operatorname{cosec} A$$

which reduces to

$$\varkappa = 0\overset{s.}{.}021 \cos z$$

or for an observation made near the almucantar passing through the pole

$$\varkappa = 0\overset{s.}{.}021 \sin\varphi$$

Since the effect of the diurnal aberration is thus shown to be constant for all stars observed at the same zenith distance, it will be most readily taken into account by applying to the clock correction derived from the uncorrected observations the correction $+\varkappa$.

The application of the preceding formulæ may be illustrated by the reduction of the following observations of four pairs of stars, made with a very small universal instrument mounted upon a portable wooden tripod. The aperture of the telescope was 33 *mm*, the magnifying power 27 diameters, the value of a level division $7''.4$:

WASHBURN OBSERVATORY, MAY 19, 1894.

COMPARISON OF CLOCK STARS WITH λ URSÆ MINORIS.

Observer, G. C. C.

Star. Circle.	ζ *Leo.* *L.*	49 *Herc.* *L.*	μ *Herc.* *R.*	ϑ *Leo.* *R.*
b_1	6.0 24.7	5.2 24.1	5.0 23.9	4.7 23.6
T_1	13 31 5	13 45 18	13 56 4	14 4 22
b_2	5.1 24.0	5.7 24.7	5.0 24.1	4.7 23.8
T_2'	13 40 43.24	13 50 4.56	14 0 1.48	14 8 19.72
Azimuth Circle	83° 35′	292° 54′	270° 18′	69° 3′
f	1.39	1.51	1.39	1.50

NOTES.— The instrument was so set that the readings of the azimuth circle are very nearly the true azimuths of the line of sight. When the vertical circle is "*Left*" the zero of the level scale is adjacent to the objective. The symbol T_2' denotes the mean of the observed times on five threads, and requires a correction to reduce it to T_2. T_1 was observed on the middle thread only. The thread intervals are assumed to be

$$I = -31\overset{s.}{.}5 \quad II = -20\overset{s.}{.}5 \quad III = 0.0 \quad IV = +20\overset{s.}{.}3 \quad V = +30\overset{s.}{.}8$$

the signs corresponding to an observation Circle R. Star W. It should be noted that owing to the reversal of the instrument the effect of any small error in the adopted thread intervals will be eliminated.

The coördinates of λ *Ursæ Minoris* and the other constants requisite for the reduction, are:

$$\alpha_1 = 19\ 29\ 25.6 \qquad log\ sin\ \pi = 8.25491$$

$$\pi = 1\ 1\ 49.88 \qquad log\ \frac{2\ sin^2\ \frac{1}{2}\ \pi}{sin\ 1''} = 1.5232$$

$$\varphi = 43\ 5\ (assumed) \qquad log\ 2\ cosec\ 2\ \varphi = 0.3020$$

Hourly rate of chronometer $= -0\overset{s.}{.}04$

Star.	*ζ Leo.*		*49 Herc.*		*μ Herc.*		*ϑ Leo.*	
	s.	s.	s.	s.	s.	s.	s.	s.
ed'n to Middle Thread	+0.25	+0.00	−0.27	−0.02	+0.25	+0.00	−0.27	+0.02
Chron. Rate. Level	+0.01	+0.54	+0.01	+0.40	0.00	−0.07	−0.01	+0.07
T_2	13 40 44.04		13 50 4.68		14 0 1.65		14 8 19.53	
α_2	10 10 49.05		16 47 17.55		17 42 20.71		11 8 42.21	
$\alpha_2 - T_2$	−3 29 54.99		+2 57 12.87		+3 42 19.06		−2 59 37.32	
$\alpha_1 - T_1$	+5 58 20.6		+5 44 7.6		+5 33 21 6		+5 25 3.6	
t	217 56 6		318 16 20		332 14 22		233 49 47	
p	66 3 13.9		74 51 7.6		62 13 18.7		73 59 31.1	
log sin p	9.96091		9.98464		9.94682		9.98283	
log cos t	9.89692 n		9.87292		9.94689		9.77099 n	
log a	8.29400		8.27027		8.30809		8.27208	
log sin t	9.78871 n		9.82321 n		9.66818 n		9.90702 n	
co log (1 − a cost)	9.9933111		0.0060815		0.0078831		9.9952305	
log cos M	9.9999692		9.9999657		9.9999798		9.9999516	
M	−0 40 57.1		−0 43 13.9		−0 33 8.7		−0 51 21.2	
x	51.3		43.6		55.3		44.3	
log tan $\frac{1}{2}(p - x)$	9.8128460		9.8837244		9.7805448		9.8769541	
$\left\{\cos M \div (1 - a \cos t)\right\}$	9.9932803		0.0060472		0.0078629		9.9951821	
log cos N	9.7770484		9.8606937		9.7593298		9.8430583	
N	+53 14 20.4		−43 28 54.2		−54 55 54.1		+45 50 8.0	
$M + N$	+ 3 30 13.55		− 2 56 48.54		− 3 41 56.19		+ 2 59 55.12	
log cot N	9.8733		0.0230 n		9.8463 n		9.9873	
C	+1.497		−2.113		−1.407		+1.946	
	s.		s.		s.		s.	
$\Delta T + C\,\Delta\varphi$	+18.56		+24.33		+22.87		+17.80	
$C\,\Delta\varphi$	− 2.32		+ 3.27		+ 2.18		− 3.02	
ΔT	+20.88		+21.06		+20.69		+20.82	

From the equations

$$
\begin{aligned}
\Delta T + 1.50\,\Delta\varphi &= 18.56 & v &= +0.02 \\
\Delta T - 2.11\,\Delta\varphi &= 24.33 & &+\ .20 \\
\Delta T - 1.41\,\Delta\varphi &= 22.87 & &-\ .17 \\
\Delta T + 1.95\,\Delta\varphi &= 17.80 & &-\ .04
\end{aligned}
$$

we obtain

$$\Delta T = +\overset{s.}{20.86} \qquad \Delta\varphi = -\overset{s.}{1.55} = -\overset{\prime\prime}{23.3}$$

with the residuals placed opposite the several equations.

In the above reduction the computations have been carried to tenths of a second of arc and hundredths of a second of time, but it is evident that quantities of this order are imperceptible in so small an instrument, a second of arc being approximately the limit of what can be seen in its telescope or measured by its level. The internal agreement of the observations as shown by the residuals is, therefore, satisfactory, and the absolute values of the latitude and clock correction furnished by the observations are also in excellent agreement with the data furnished by a geodetic connection with the Washburn Observatory and a comparison of the chronometer with the normal clock. Thus after correcting ΔT for diurnal aberration we have

	s.	° ′ ″
From Observation	$\Delta T = +20.88$	$\varphi = 43\ 4\ 36.7$
From Comparisons	$= +20.80$	$= 43\ 4\ 36.5$

This excellent agreement is due, at least in part, to the reversal of the instrument, one-half of the observations having been made Circle Right and one-half Circle Left, thus eliminating the effect of error in the assumed thread intervals.

In order to secure the convenient observation of stars it will be advantageous to prepare in advance an observing programme showing the time at which the several clock stars cross the almucantar of the polar star, and their corresponding azimuths. If only a few stars are to be included in the programme this can be most conveniently done by putting $T_2 = T_1$ in equations (13) and solving with

four place logarithms the following approximate equivalents of those equations:

$$t = \alpha_2 - \alpha_1 \qquad \tan M = \frac{a \sin t}{1 - a \cos t}$$

$$a = \sin \pi \operatorname{cosec} p \qquad \cos N = \frac{\tan \varphi \tan \frac{1}{2} p}{1 - a \cos t} \tag{18}$$

$$T_1 = T_2 = \alpha_2 - \Delta T + M + N$$

When the sidereal times T_1 and T_2 are known the zenith distances and azimuths of the stars may be directly computed from the fundamental formulæ for the transformation of coördinates, but the following method will usually be found more convenient:

In the spherical triangle formed by the polar star, the zenith and the pole, we represent the east hour angle of the star by τ and find

$$\begin{aligned} \cos z &= \sin \varphi \sin \delta_1 + \cos \varphi \cos \delta_1 \cos \tau \\ &= \cos (\delta_1 - \varphi) - \cos \varphi \cos \delta_1 \, 2 \sin^2 \tfrac{1}{2} \tau \end{aligned}$$

and applying to this the development into series of

$$\cos x = \cos y + h$$

find when terms of the order π^2 are neglected

$$z = H - \varphi \qquad H = 90^\circ - \pi \cos \tau \tag{19}$$

Similarly from the development of the azimuth into series we find when the azimuth is reckoned from the north, positive toward east,

$$A_1 = \pi \sin \tau \sec \varphi = M_0 \sec \varphi \tag{20}$$

Values of H and M_0 with the argument τ are tabulated below.

To determine the difference of azimuth of the stars, we represent by ρ the length of an arc of a great circle joining them, and from the isosceles spherical triangle formed by the two stars and the zenith, find

$$\cos \rho = \cos^2 z + \sin^2 z \cos (A_2 - A_1)$$

which is readily transposed into either

$$\sin \tfrac{1}{2} (A_2 - A_1) = \sin \tfrac{1}{2} \rho \operatorname{cosec} z$$

or

$$\tan \tfrac{1}{2} (A_2 - A_1) = \frac{\sin \frac{1}{2} \rho}{\sqrt{\sin \left(z - \frac{\rho}{2}\right) \sin \left(z + \frac{\rho}{2}\right)}} \tag{21}$$

The first of these equations will usually be the more convenient.

To determine ρ we have from the triangle formed by the two stars and the pole

$$\cos\rho = \cos\pi \sin\delta + \sin\pi \cos\delta \cos(\alpha_2 - \alpha_1)$$

where δ is the declination of the southern star. In place of this rigorous equation we may write with sufficient precision

$$\rho = 90 - \delta - \pi \cos(\alpha_2 - \alpha_1) = H(t) - \delta \qquad (22)$$

where the symbol $H(t)$ denotes the tabular value of H corresponding to the argument $t = \alpha_2 - \alpha_1$.

Equations (18), (20) and (22), in connection with the tabular values of H and M_0, suffice for the construction of an observing list, but if any considerable number of stars are to be observed in the same latitude it will be found an economy of labor to construct for the given latitude special tables, such as those given below for the Washburn Observatory, which are based on the following analysis:

Neglecting terms of the order π^2 we put

$$\cos N_0 = \tan\varphi \tan\tfrac{1}{2}p$$

and find from equations (18)

$$\cos N = \cos N_0 + a \cos t \cos N_0$$
$$N = N_0 - \pi \cos t \sec\delta \cot N_0$$
$$M = \pi \sin t \sec\delta$$

The factor $\pi \sin t = M_0$ has been tabulated, and $-\pi \cos t$ is evidently equal to the tabular value of M_0 which corresponds to the argument $t - 6^h$. Putting

$$\sec\delta = h \qquad \sec\delta \cot N_0 = k$$

we tabulate N_0, h and k with the argument δ and find for the instant at which the two stars have equal altitudes

$$T = \alpha + N_0 + h M'_0 + k M''_0$$

the accents ′ ″ denoting that the arguments for the corresponding values of M_0 are t and $t - 6^h$.

It should be noted that since N_0 is an approximation to the N of the rigorous formulæ we shall have N_0 and k positive for a star west of the meridian and negative for a star east of the meridian.

Similar tables may be constructed for the difference of azimuth of the stars, but the direct computation by (19) and (21) is so simple that little advantage would be derived from them.

To illustrate the use of the tables we make the following computations for a comparison of Polaris with ρ Leonis west of the meridian and δ Herculis east of the meridian:

$$\varphi = 43^\circ \; 4'.6 \qquad \alpha_1 = 1^h \; 18^m.9$$

Star.	*ρ Leonis.*	*δ Herculis.*
α_2	10 27.3	17 10.7
t	9 8.4	15 51.8
δ	9 51	24 58
$H(t)$	90 54	90 39
ρ	81 3	65 41
N_0	+ 2 32.5	− 3 33.7
$h\,M_0'$	+ 3.4	− 4.6
$k\,M_0''$	+ 4.6	− 2.1
T	13 7.8	13 30.3
$H(\tau)$	91 14	91 14
z	48 10	48 10
cosec z	0.1278	0.1278
$\sin \frac{\rho}{2}$	9.8127	9.7343
$A_2 - A_1$	121 23	93 25

The azimuth of Polaris corresponding to the times above computed is, in minutes of arc, $15 \sec \varphi \cdot M_0'$.

For a determination of time only, the latitude being supposed known or not required, the observation of the polar star may be omitted and the observation confined to noting the times at which two southern stars reach the same altitude. Convenient formulæ and tables for observations of this kind have been published by Wittram.[1]

AUXILIARY TABLES FOR TRANSITS OVER THE ALMUCANTAR OF POLARIS.

FOR ALL LATITUDES.

t or τ	M_0		H		t or τ
h.	m.		° ′		h.
0	+ 0.0 −		88 46		24
		1.3		3	
1	+ 1.3 −		88 49		23
		1.2		7	
2	+ 2.5 −		88 56		22
		1.0		12	
3	+ 3.5 −		89 8		21
		0.8		15	
4	+ 4.3 −		89 23		20
		0.5		18	
5	+ 4.8 −		89 41		19
		0.1		19	
6	+ 4.9 −		90 0		18
		0.1		19	
7	+ 4.8 −		90 19		17
		0.5		18	
8	+ 4.3 −		90 37		16
		0.8		15	
9	+ 3.5 −		90 52		15
		1.0		12	
10	+ 2.5 −		91 4		14
		1.2		7	
11	+ 1.3 −		91 11		13
		1.3		3	
12	+ 0 0 −		91 14		12

[1] Tables Auxiliaires pour la Détermination de l'Heure par des Hauteurs Correspondantes de Différentes Etoiles. St. Petersburg, 1892.

FOR THE LATITUDE 43° 4.′6

δ	N_0 +W.	−E.	h	k +W.	−E.	δ
°	h. m.					°
−2	0 57.9	13.8	1.00	3.88	79	−2
−1	1 11.7	11.4	1.00	3.09	45	−1
0	1 23.1	9.9	1.00	2.64	31	0
+1	1 33.0	8.9	1.00	2.33	23	+1
2	1 41.9	8.0	1.00	2.10	17	2
3	1 49.9	7.3	1.00	1.93	15	3
4	1 57.2	6.9	1.00	1.78	11	4
5	2 4.1	6.6	1.00	1.67	10	5
6	2 10.7	6.1	1.01	1 57	9	6
7	2 16 8	5.7	1.01	1.48	7	7
8	2 22.5	5.5	1.01	1.41	7	8
9	2 28.0	5.3	1.01	1.34	6	9
* 10 *	2 33.3	* * 9.8	* 1.02 *	1.28	* * 10	* 10 *
12	2 43.1	9.2	1.02	1.18	8	12
14	2 52.3	8.5	1.03	1.10	7	14
16	3 0.8	8.0	1.04	1.03	6	16
18	3 8.8	7.6	1.05	0.97	5	18
20	3 16.4	7.2	1.06	0.92	4	20
22	3 23.6	6.9	1.08	0.88	4	22
24	3 30.5	6.5	1.09	0.84	4	24
26	3 37.0	6.3	1.11	0.80	3	26
28	3 43.3	6.0	1.13	0.77	3	28
30	3 49.3	5.8	1.15	0.74	3	30
32	3 55.1	5.7	1.18	0.71	2	32
34	4 0.8	5.4	1.21	0.69	2	34
36	4 6.2	5.3	1.24	0.67	2	36
38	4 11.5	5.1	1.27	0.65	2	38
40	4 16.6		1.31	0.63		40

www.ingramcontent.com/pod-product-compliance
Lightning Source LLC
Chambersburg PA
CBHW021642270326
41931CB00008B/1126
* 9 7 8 3 3 3 7 3 9 5 9 4 0 *